The Little Lobster

A LOBSTER TALE

By W. Thomas Hotz

Illustrated by Estelle Corke

STAR BRIGHT BOOKS

CAMBRIDGE MASSACHUSETTS

Published in the United States of America by Star Bright Books, Inc.
The name Star Bright Books and the Star Bright Books logo are registered trademarks
of Star Bright Books, Inc. Please visit: www.starbrightbooks.com.
For orders, email: orders@starbrightbooks.com or call: (617) 354-1300.

Hardcover ISBN: 978-1-64909-080-5
Paperback ISBN: 978-1-64909-081-2
Ebook ISBN: 978-1-64909-082-9
Star Bright Books / MA / 00105250
Printed in China / WKT / 10 9 8 7 6 5 4 3 2 1

Printed on paper from sustainable forests.

Library of Congress Cataloging-in-Publication Data is available.

In memory of W. Thomas Hotz whose love and respect for the ocean and all sea creatures large and small will continue to live on in the pages of this book. —Mary Jo Lagana

For Lucy —Estelle Corke

Deep down in the North Atlantic Ocean, off the rocky coast of Maine, a mother lobster crawls and glides on the tips of her slender walking legs through the swaying sea grass on the ocean floor.

Nestled underneath her tail are thousands of baby lobster eggs. She carries them attached by a sticky substance to the hairs on her swimmerets, a double row of swimming paddles under her tail. Her eggs are black and about the size of a pinhead. As they develop, they change from black to green and then to golden brown and will even have blue eyes!

A view from above

A view from underneath

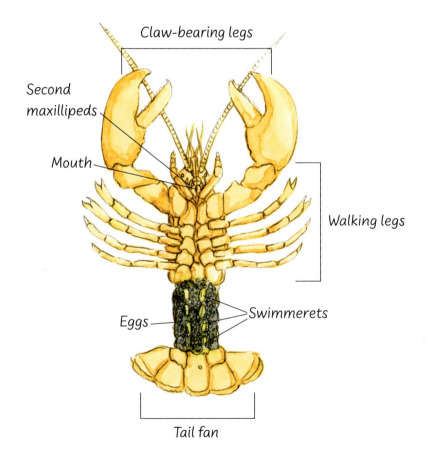

Crusher claw

Antenna

Antennulas

Eyes

Pincer claw

Carapace
(body shell from
eyes to start of tail)

Abdomen
(or called Tail)

Outer tail fins

Central tail fin

Claw-bearing legs

Second
maxillipeds

Mouth

Walking legs

Eggs

Swimmerets

Tail fan

The mother lobster is very careful with her eggs because it will be ten to eleven months before they hatch. During this period, from time to time, she will flutter her swimmerets back and forth to keep the eggs healthy and clean.

She will also pick out the sick and damaged eggs with her hind legs. This is called preening, and she loses nearly ten to fifteen percent of her eggs in the process so that the rest will have a chance to survive. Also, when she senses danger, she curves her hard-shelled tail around the cluster of eggs to shield them.

All of this activity makes the mother lobster very hungry, and as she searches the ocean floor under the cover of darkness, her sensitive antennae pick up the distinct scent of fish.

This leads her to a wire basket with chunks of fish cradled and hanging inside. She climbs all over this strange object, her excited antennae waving faster and faster until she finds an opening and drops in.

Inside, she picks at the fish with her second pair of appendages called maxillipeds. They act like small hands that feed pieces of fish into her mouth. After eating most of this easy meal, the mother lobster tries to leave, but there is no escape!

In the early morning there is a tug and suddenly the cage quickly lifts out of the water. The mother lobster and all the attached eggs have been caught!

The lobster fisherman is happy to see such a large lobster in his trap. But when he looks underneath and sees all the eggs, he realizes that she is a berried or egger lobster, and it would be illegal to keep her.

The lobster fisherman gently puts her back in the water, allowing for the possibility that the eggs will hatch and add to the future number of lobsters.

Now, after all these months the mother lobster is ready to hatch her eggs. But she does not bury the eggs in the sand like other marine animals. Instead, during the night she raises her tail and rapidly beats her swimmerets. From under their mother's tail, thousands of baby lobsters stream out into the open ocean, each one shedding its casings and floating aimlessly to the surface.

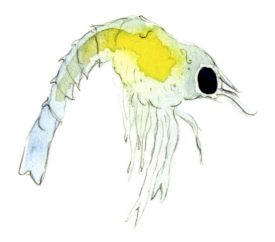

First larval stage after being released from the mother. Completely transparent.

In the second larval stage, the lobster has started to develop its swimmerets but they are not fuctional.

The third larval stage. Slightly larger than stage two, but still can't use its swimmerets. The lobster uses its front appendages to swim.

The newly hatched lobsters do not resemble their mother. Instead, they look like shrimps with large black eyes and hairy-feathery legs. They are now called larval lobsters. Larval lobsters are transparent, so their organs are visible through the shell.

After being released from their mother's tail, the larval lobsters go through three larval growth stages in their first month of life followed by a post-larval stage.

At the fourth stage of growth many changes occur. Unlike the first three stages when larval lobsters looked like small insects, they have now developed tiny claws, a powerful tail, and long antennae.

In the post-larval stage, lobsters begin to look more recognizable. Their body slightly straightens, their claws are stretched in front of them, and they use their swimmerets to swim.

Carried by currents and drifting up and down in the water column, they are exposed to many dangers. Fish, floating jellyfish, surface-feeding seabirds, and even other larval lobsters devour them, leaving only about ten to survive out of more than ten thousand eggs that the mother lobster hatched.

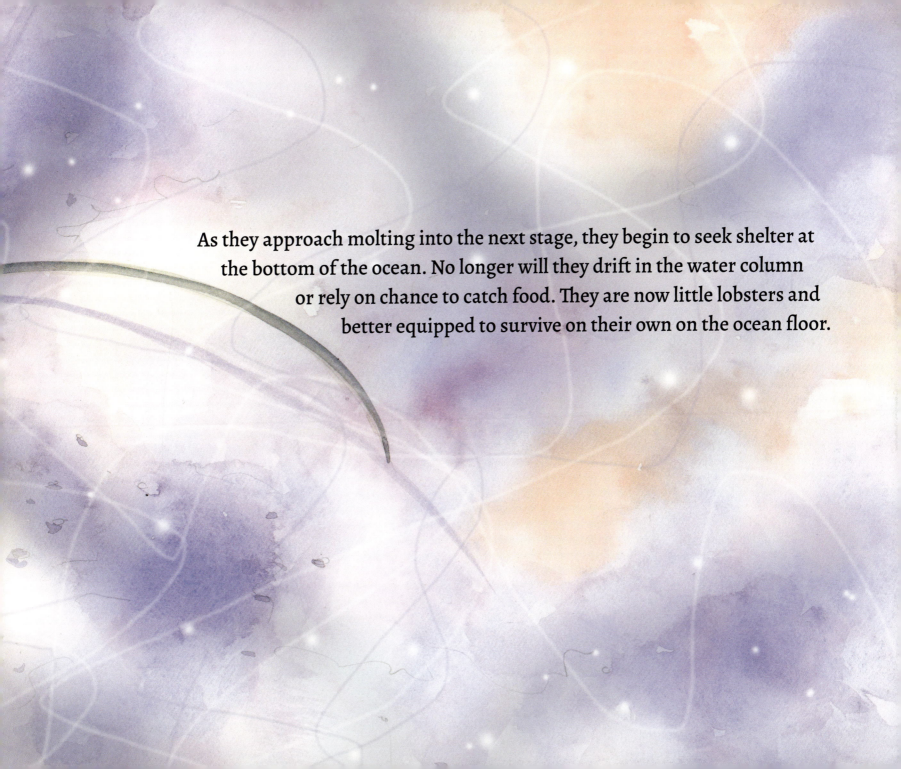

As they approach molting into the next stage, they begin to seek shelter at the bottom of the ocean. No longer will they drift in the water column or rely on chance to catch food. They are now little lobsters and better equipped to survive on their own on the ocean floor.

But life on the ocean floor is very dangerous. Fish, crabs, and starfish, all hunting for food, can easily consume them. Lobsters at this time tend to separate from each other to find a place to hide.

A little lobster, using its antennae and the sensitive hairs on its shell, "feels" the vibrations in the water to find a tiny crack in some rocks.

The little lobster retreats into the small space and, with its claws and antennae at the entrance to its shelter, captures any tiny ocean creature that swims by.

The little lobster does not spend all its time there. Under the protection of darkness, it ventures out to explore the area, always hunting for food.

Soon after the little lobster begins living on the ocean floor, it becomes restless and very uncomfortable. Its body starts to swell up from the inside of its shell. Since the shell does not grow along with the body inside, the shell splits apart right down the back. The little lobster struggles and finally wiggles out. A new, larger shell has formed underneath, but it is very soft.

Over the next several days, the little lobster takes in great amounts of water and eats its old shell to fill up to its new size. It has just gone through its first molt and will undergo seven to ten more molts in its first year, doubling in size and weight each time.

Finally, little lobster will be a juvenile lobster, ready to take on the challenge of growing up to be an adult!

Juvenile lobsters can molt up to 25 times in their first 5 to 7 years before reaching adulthood. This means it takes 5 to 7 years for a lobster to grow to the legal size for harvesting, which is approximately 1 pound. Adult male lobsters typically molt once a year, while females molt once every two years. After a female lobster has molted and is in the soft-shell state, she mates and begins a new life cycle.

A Little Lobster History

Centuries ago, before the United States was colonized, Native Americans used lobsters in many ways. Not only would they eat the shellfish, but they were used as fishing bait and fertilizer. When the European settlers landed ashore in the Northeastern US, it is said that there were so many lobsters in the ocean that they were washing up on shore. Because they were so common lobsters didn't cost much, and they became known as the poor man's protein. As train travel expanded, many people who had never tried lobster now had the chance, and it soon became a delicacy and sold at high prices. This eventually led to overfishing.

Lobsters are bottom-dwellers, which means they live at the bottom of the ocean. They thrive in cold water, from 59-64°F (15-18°C), and are found in oceans around the world from Great Britain to South Africa and western US, but the American lobster is most commonly found in the Atlantic coast along the eastern US. Because of climate change, the warming oceans have caused the American lobster population to shift northward, making the Gulf of Maine home to the largest lobster population.

Today, the lobster industry and lobstermen are more aware of climate change and the protection of juvenile lobsters. There are now rules in place about how lobster-men can catch these shellfish. In order to protect these animals, allow them to grow larger, and let them reproduce more, the caught lobsters' carapace must be between 3.25 and 5 inches long, and they must release female lobsters who are carrying eggs. Some states, like Maine, have more rules, like using trap vents that allow smaller caught lobsters to escape and making a V notch on females' shells to show other lobstermen that this lobster is egg-bearing and to release it if caught.

Lobsters aren't the only marine life being affected by the changing climate and overfishing. As of 2024, only about 350 North Atlantic right whales, a critically endangered species, remain. Many of these whales show signs of getting caught in ropes. Environmental groups think this might be from the lobstermen's traps, and while lobster-men understand that using less rope by attaching more traps onto one rope might help, it would make their job more dangerous. Lobstermen also want to do their part to protect lobsters and marine life. Both lobstermen and environmental groups will continue to do what they can to ensure that lobsters, whales, and other marine life are protected.

Glossary

Appendage: An appendage is an outer body part that sticks out from the body. On humans, arms, legs, or fingers are examples of appendages. On a lobster, its legs, swimmerets, maxillipeds, or claws are all appendages.

Berried lobster: A berried lobster, or egger lobster, is a female lobster who is carrying eggs. A female lobster carries the eggs under her tale and can carry 8,000 to 100,000 eggs depending on her size.

Crusher claw and pincer claw: A lobster has two claws. The larger claw is called the crusher claw and is used to crush the food it finds. The smaller claw is called the pincer claw and is used to tear its prey apart in order to eat it easier. Some lobsters have the crusher claw on the right and the pincher claw on the left, while others have the opposite.

Current: A current is the movement of water. Water currents happen in any body of water like the ocean or rivers. The water is moved by wind, gravity, temperature, and the rotation of the Earth.

Larval lobster: When lobsters first hatch from their eggs they float and move toward the ocean surface. Lobsters in this stage are larval lobsters. They stay in this phase for 3-10 weeks. During these weeks, the larval lobster will molt four times, each time getting bigger and developing its appendages.

Maxilliped: The maxillipeds are the mouth parts of the lobster. They are also called "jaw legs" and are used to pass food to the mandible, or the chewing, jaw-like part of the lobster.

Molting: As lobsters grow, they begin to outgrow their shell. The lobster drinks and absorbs a lot of water. This water intake causes the new shell to swell, pushing the old outer shell until it splits. The lobster pulls itself out with a new shell. The whole process is called molting, and lobsters molt many times throughout their lives.

Ocean floor: The very bottom of the ocean is called the ocean floor, or the seabed, and this is where lobsters live. The ocean floor covers more than 70% of our world's surface, and it has many features of dry land, like mountains, canyons, and even volcanoes.

Preening: When a berried lobster uses its hind legs to keep her eggs clean and tidy, this is called preening. She will use her hind legs to get rid of the sick or damaged eggs to keep the other eggs healthy.

Swimmerets: Swimmerets are small, skinny fins on the underside of a lobster's tail. They might look like little legs, but they are actually used to help lobsters swim forward.

Water column: A water column is a vertical expanse of water stretching between the ocean surface and the ocean floor. When someone says the water column, they could be talking about a very specific area of the ocean or the entire thing.